Space

**Written by
Nigel Nelson
Illustrated by
Rebecca Archer**

Wayland

Topic Web

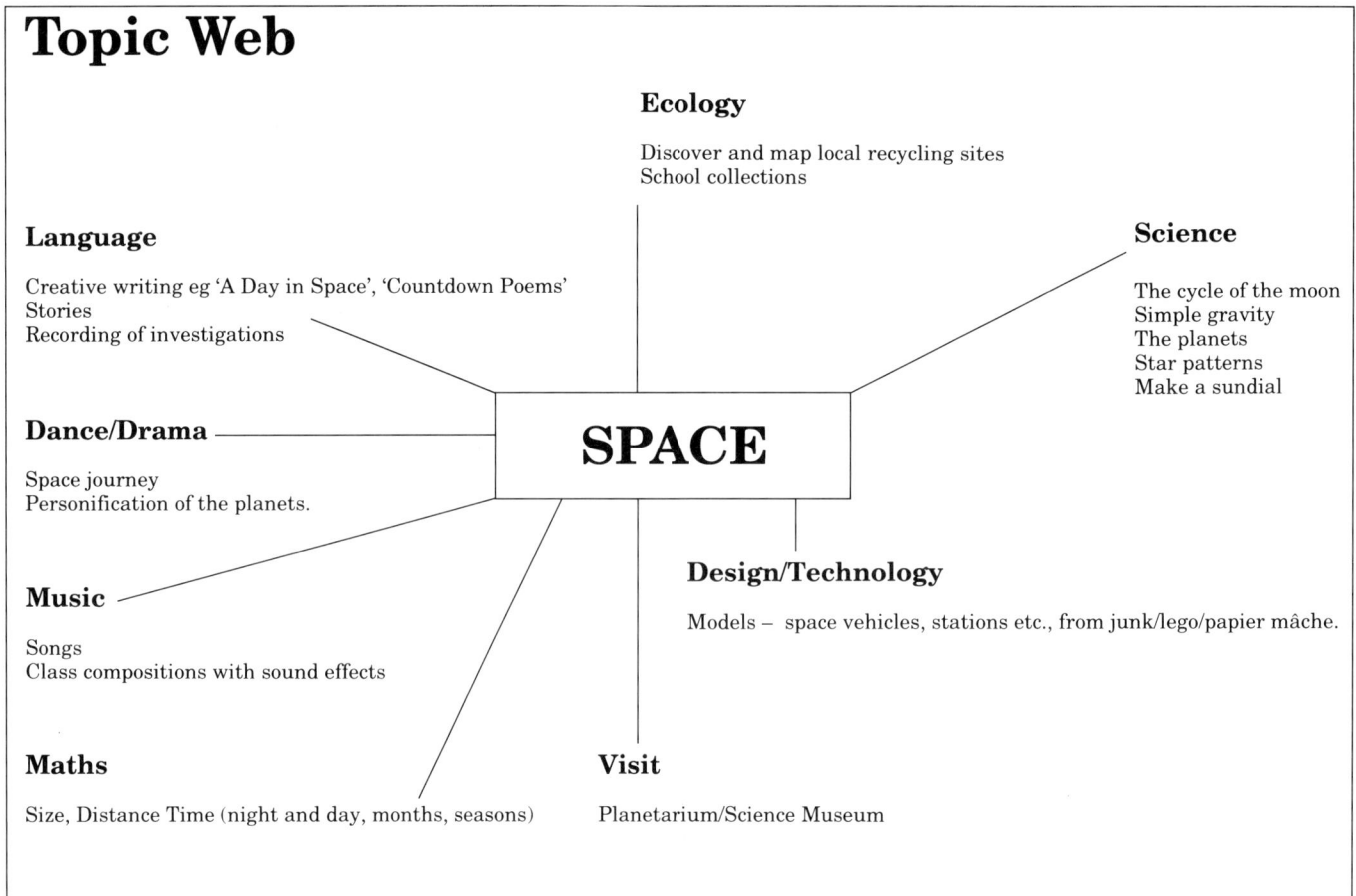

Ecology

Discover and map local recycling sites
School collections

Language

Creative writing eg 'A Day in Space', 'Countdown Poems'
Stories
Recording of investigations

Science

The cycle of the moon
Simple gravity
The planets
Star patterns
Make a sundial

Dance/Drama

Space journey
Personification of the planets.

SPACE

Music

Songs
Class compositions with sound effects

Design/Technology

Models – space vehicles, stations etc., from junk/lego/papier mâche.

Maths

Size, Distance Time (night and day, months, seasons)

Visit

Planetarium/Science Museum

First published in 1992 by
Wayland (Publishers) Ltd
61 Western Road, Hove
East Sussex, BN3 1JD, England

© Copyright 1992 Wayland (Publishers) Ltd

British Library Cataloguing in Publication Data

Nelson, Nigel
Space
1. Title
523

HARDBACK ISBN 0 7502 0605 5

PAPERBACK ISBN 0 7502 0803 1

Editor: Mandy Suhr
Designer: Loraine Hayes
Consultant: Jane Battell

Typeset by Dorchester Typesetting Group Ltd
Printed and bound by Casterman S.A. in Belgium

Picture Acknowledgements
The pictures in this book are by: Bruce Coleman (Jeff Foott) 29; Genesis Space Photo Library 10, 16 (NASA), 17 (NASA), 25 (below), 27; Science Photo Library cover, 5 (NASA), 14 (NASA), 18 (NASA), 25 (NASA), 26 (Chris Bjornberg), 28 (Sinclair Stammers); TRH/NASM/NASA 7.

Contents

The words that appear in **bold** are explained in the glossary.

Planet Earth

Our **planet** Earth is a huge rocky ball. From space it looks like a blue planet wrapped in white clouds. This is because most of the Earth is covered with water. Around the Earth is a thin blanket of air. We call this the **atmosphere**.

The atmosphere is a mixture of **gases**. It keeps the Earth warm by trapping heat from the Sun. It also protects our planet from the Sun's harmful rays.

Into Space

When you jump up in the air, you are pulled down to the ground again by gravity. This invisible force holds everything on Earth in place by gently pulling it towards the centre of the Earth. All the planets have their own gravity.

Gravity makes it very difficult to go into space. Powerful rocket engines are needed to **launch** spaceships. These are stronger than the Earth's gravity and send the rocket into space.

In a rocket, you soon leave the Earth's atmosphere and enter space. There is no air to breathe and it is dark and cold.

The Moon

The Moon is the nearest thing to us in space. It is four times smaller than the Earth. The Moon travels round and round the Earth in an **orbit**. Each orbit takes about one month.

The Moon doesn't have its own light.
You can see the moon because the
light from the Sun shines on it. There
is no air or water on the Moon.

The **craters** on the Moon
were probably caused by
rocks from space crashing
into it.

Walking on the Moon

Before people went to the Moon, space **probes** were sent there to check if it was safe to land. Scientists discovered that **astronauts** would need to wear special spacesuits to go there.

On the Moon you can jump a lot higher than on Earth, and you can lift heavy rocks. This is because the Moon's gravity is not as strong as the Earth's gravity.

The first person to walk on the Moon was an American called Neil Armstrong, in 1969.

Space Travel

Sputnik 1 was the first satellite to be launched into space in 1957.

The first person to travel in space was a Russian called Yuri Gagarin, in 1961.

The first space traveller was a Russian dog called Laika.

12

The Apollo spacecraft from the USA made trips to the Moon.

The Space Shuttle can be launched over and over again.

Spacesuit

Although we can't see or feel it, air is all around us. We need the oxygen gas in air to breathe. There is no air in space so astronauts have to take oxygen with them in their backpacks.

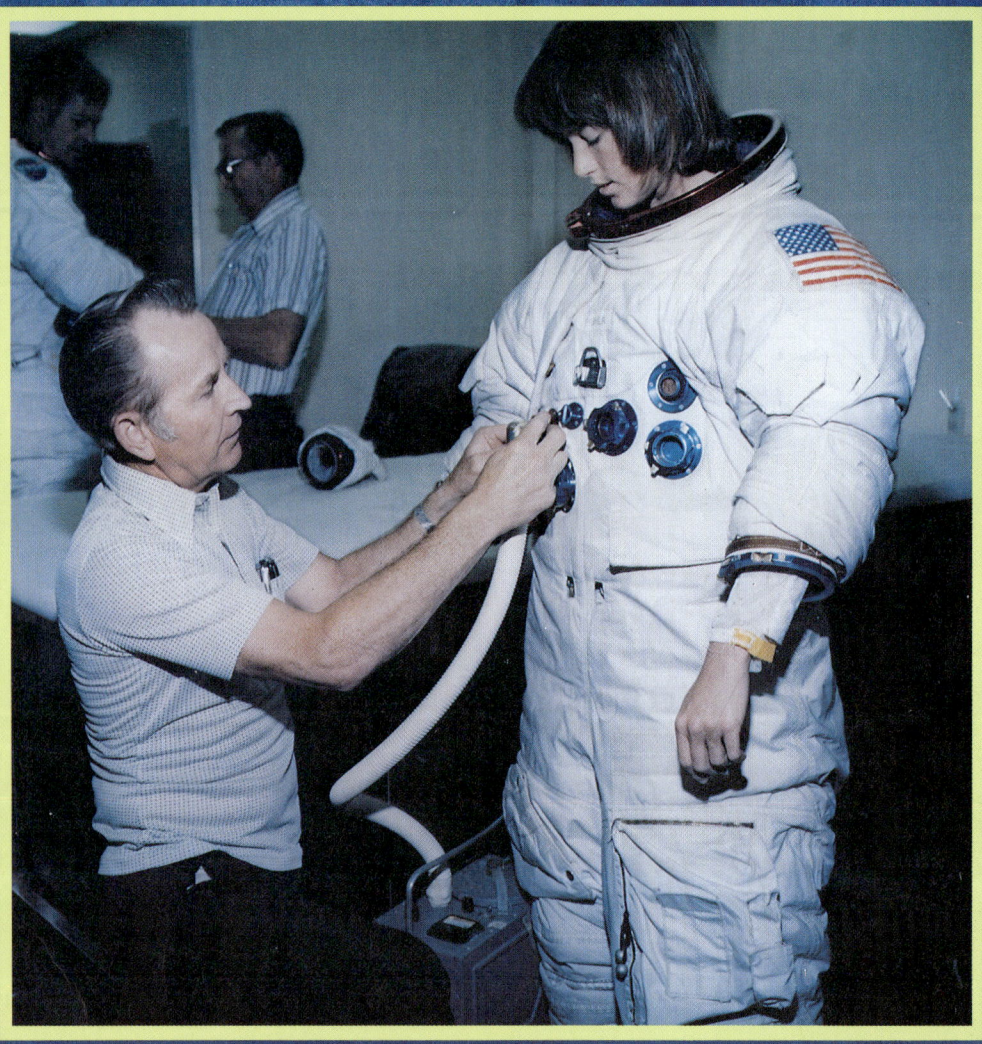

A spacesuit is very heavy on Earth but feels much lighter in space.

The helmet has a dark **visor** to keep out the Sun's harmful rays.

As well as oxygen, water is carried in the backpack.

Even though it is cold in space, it gets hot inside the suit. Special water tubes inside keep it cool.

Living in Space

People can live in space stations for months and months. Spacesuits are not needed because the station is filled with air to breathe. But in space there is no gravity to pull you downwards and so you float about.

Food and drink will float around so space food is extra sticky. You have to drink through a straw.

You sleep in a sleeping bag which is tied down to stop you floating around.

You have to exercise or your muscles will go floppy.

The Sun

The Sun is really just a medium sized **star** – the nearest one to us. It is massive compared with Earth. The Sun is an incredibly hot burning ball of gases. Without the light and the heat it gives us there would be no life on Earth.

As the Earth moves in an orbit around the Sun it also spins round and round. It takes twenty-four hours to spin right round. It is night time when your part of the Earth is facing away from the Sun.

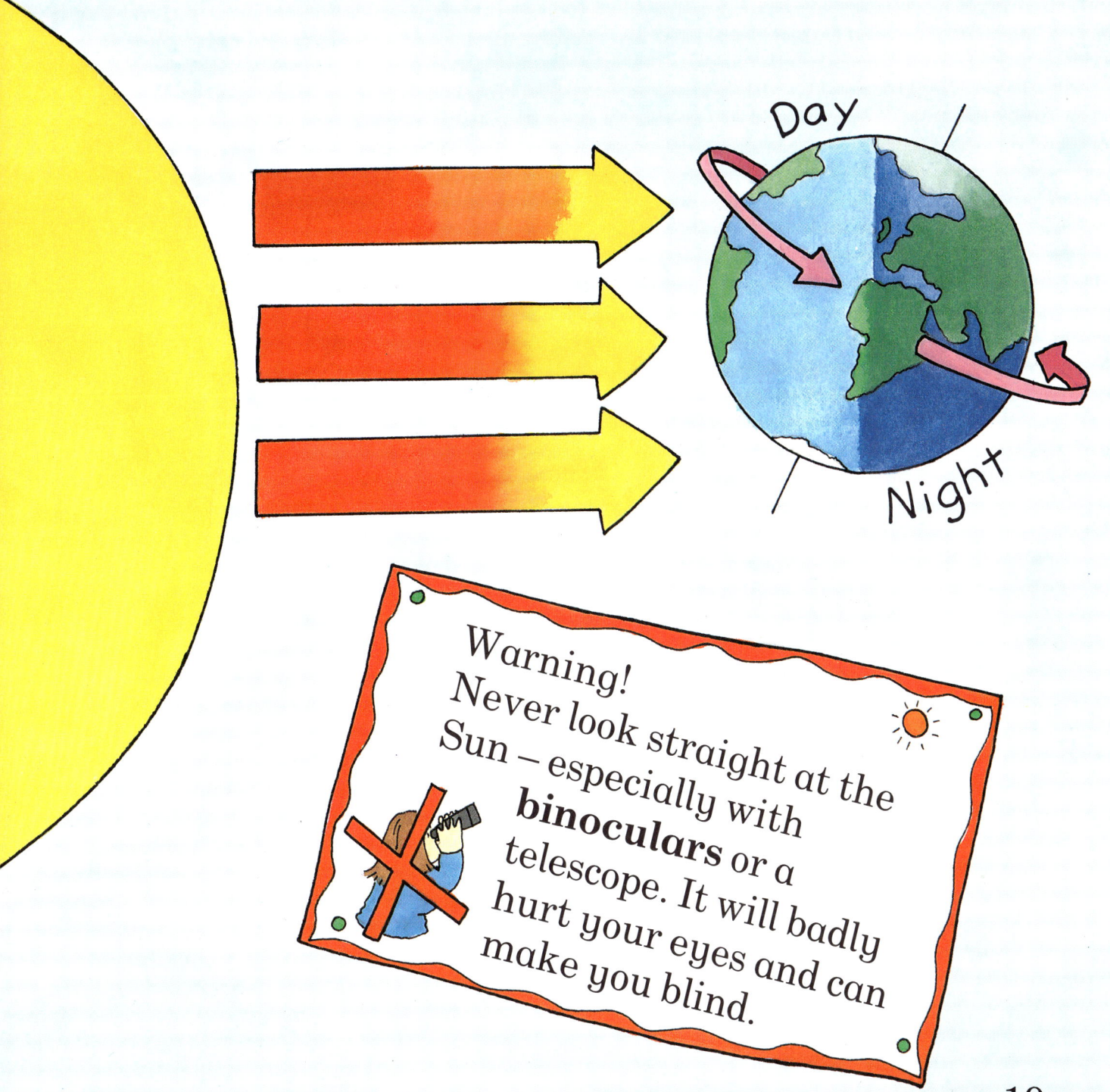

Day

Night

Warning! Never look straight at the Sun – especially with **binoculars** or a telescope. It will badly hurt your eyes and can make you blind.

The Solar System

The Earth is only one of nine planets which orbit the Sun. Can you learn their names? The planets are millions of kilometres apart.

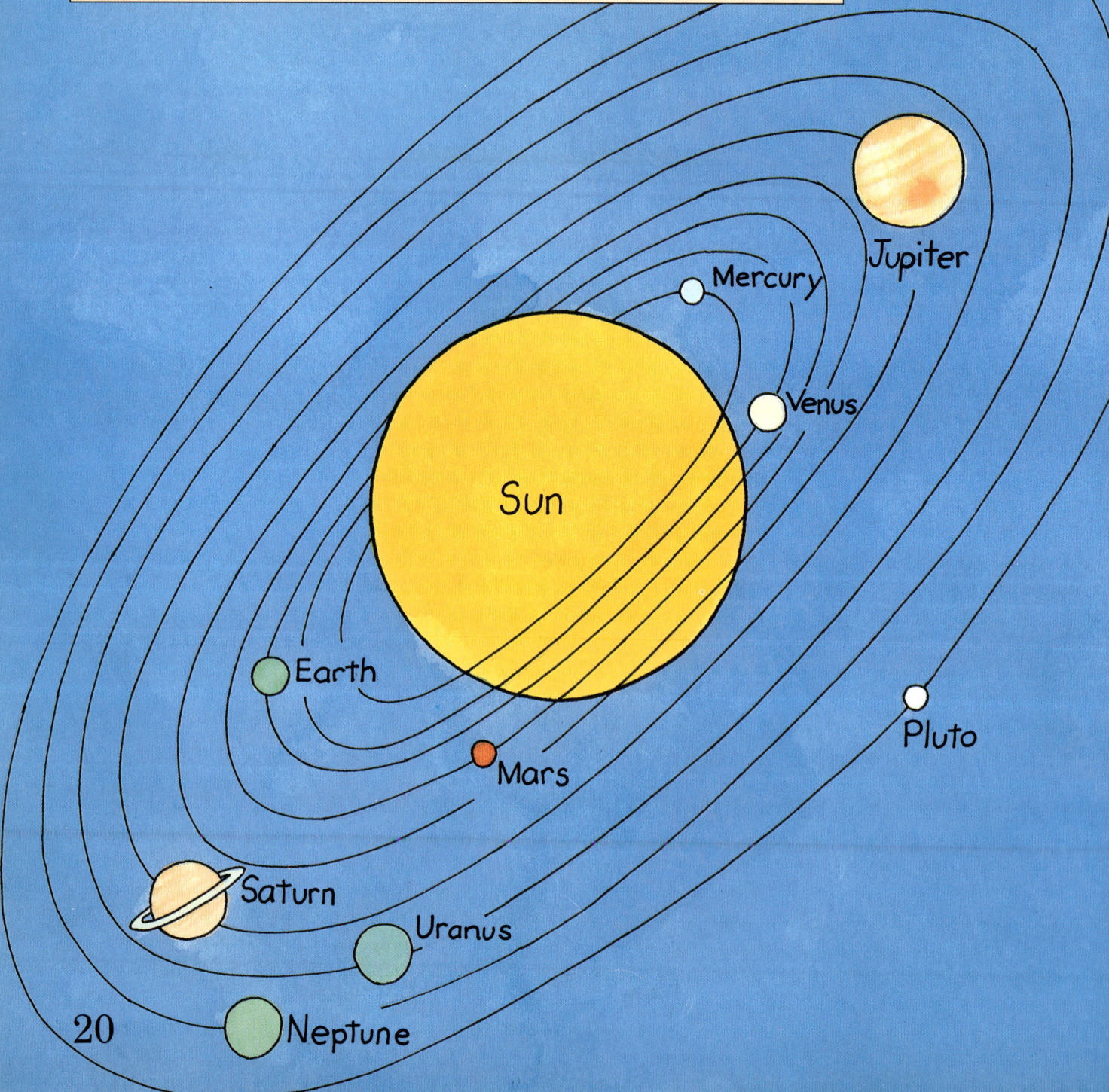

Jupiter

Mercury

Venus

Sun

Pluto

Earth

Mars

Saturn

Uranus

Neptune

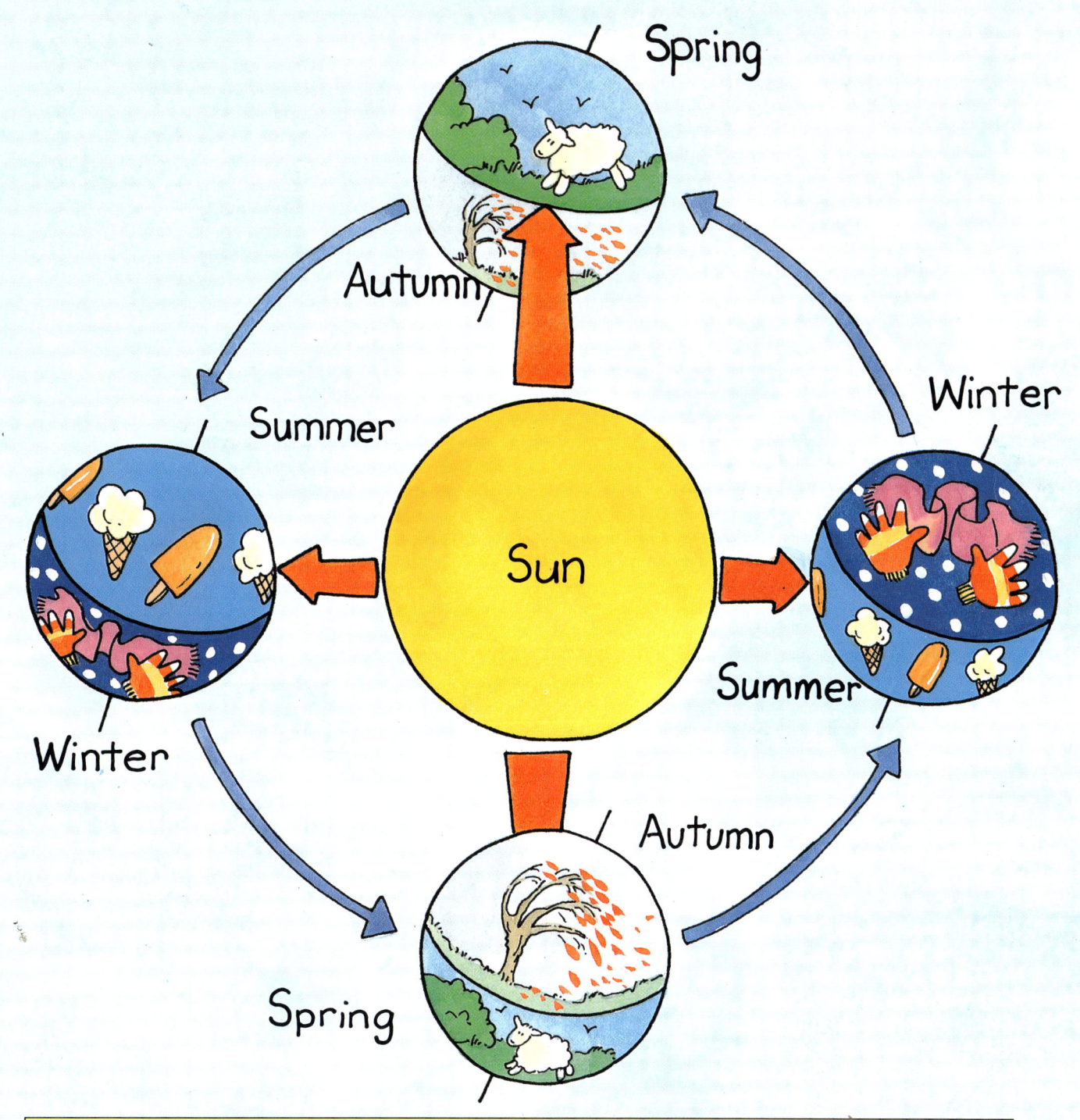

Spring

Autumn

Summer

Winter

Sun

Winter

Summer

Autumn

Spring

It takes a year for the Earth to go right around the Sun. Because the Earth is tilted, some parts of the world get more warmth from the Sun at some times of the year and less at other times. This is what gives us our seasons.

The Universe

The Earth seems big but it's only a speck in space. Our sun is just one of millions of other stars in our **galaxy** which is called the Milky Way. The Universe is made up of millions and millions of galaxies.

Look for the patterns in the sky made by the stars.
They all have different names.
Can you find this pattern? It is called the **plough**.

Earth is Special

The Earth is the only planet we know where there are living things. Because of its distance from the Sun, the Earth is not too hot and not too cold. Its atmosphere contains just the right mixture of gases needed by plants, animals and people.

24

Venus is a rocky planet like Earth but you couldn't live there. It is far too hot and its atmosphere contains clouds of burning **acid**.

Saturn is one of the **giant planets** like Jupiter, Uranus and Neptune. You couldn't even land on Saturn because it is mainly made of **liquid**.

Living on Another Planet

Mars is called the red planet because the soil and rocks are red.

Some day people might live on another planet. To live on Mars would be possible but not easy. It is much colder and all the water is frozen or underground. It has a thin atmosphere which we could not breathe and which would not protect us from the Sun's harmful rays.

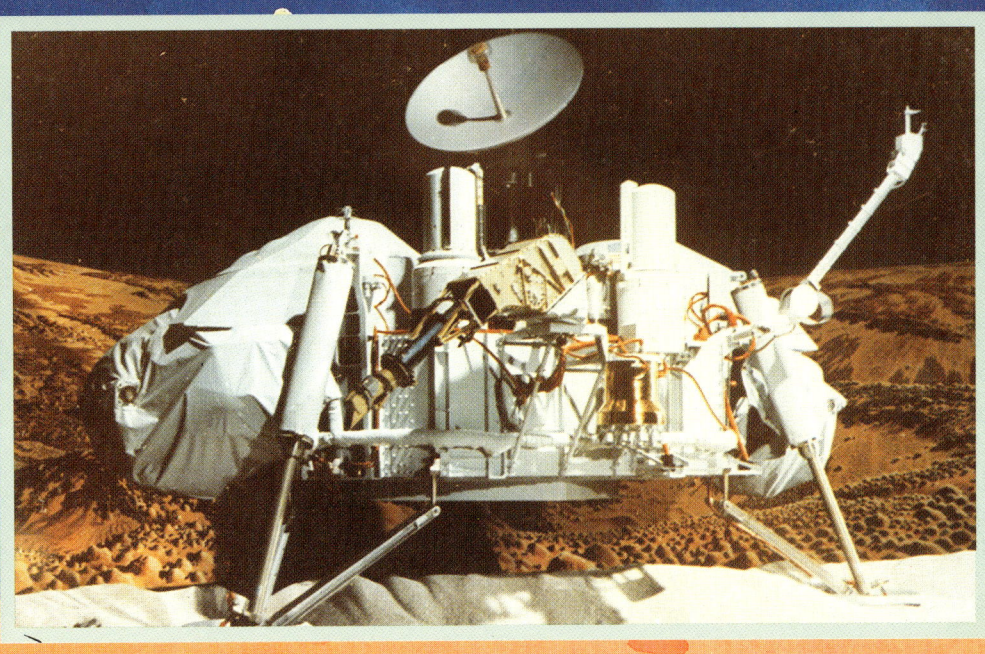

Mars is a long way away. The *Viking* spacecraft took nearly a year to journey to Mars.

Looking After Our Planet

Our planet Earth is very special. We must learn to look after it carefully. It's important to know what harms our planet and how we can help it.

The gases in some **aerosols** can harm part of the atmosphere called the **ozone layer**.

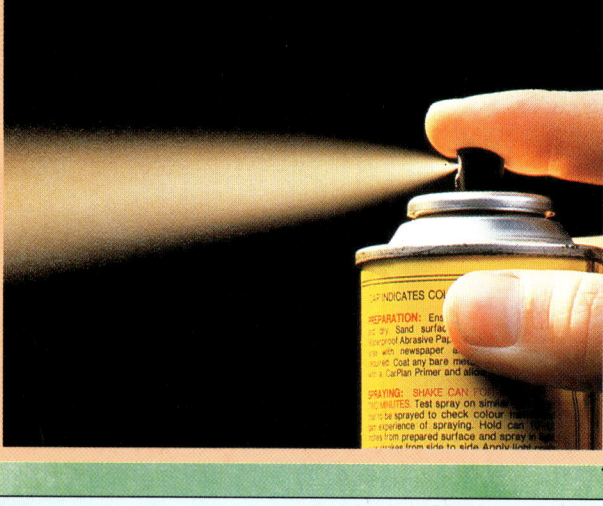

Throwing things away that can be used again is a waste of resources. **Recycle** things like paper, glass and cans.

Save electricity! Most of it is made by burning **fuels**. Some experts think that a change in the atmosphere may be caused by this burning.

29

Glossary

Acid A liquid which can burn the skin.

Aerosol A spray can.

Astronauts People who travel in space.

Atmosphere A layer of gases around a planet.

Binoculars Something which you look through with both eyes to see things far away.

Craters Bowl-shaped dents in the surface of a planet or moon.

Fuels Things that are burned to produce heat or power.

Galaxy A large group (millions and millions) of stars and planets.

Gases A substance, like air, that is not a liquid or a solid.

Giant planets The name given to the biggest planets (Jupiter, Saturn, Uranus and Neptune) in our solar system which are mainly made of gases and liquid.

Launch To send into space.

Liquid A substance, like water, that can flow.

Orbit The path something follows as it goes round and round a planet or a star.

Ozone layer An area of the atmosphere where ozone gas is present. This protects us from the Sun's harmful rays.

Planet A large sphere in space that orbits the Sun or some other star.

Plough A group (or constellation) of stars which looks like an old-fashioned plough.

Probe A small spacecraft (usually without people) used to explore space.

Recycle To change waste materials into something which can be used again.

Star A ball of burning gas in space.

Visor The part of a helmet which protects the eyes.

Books to Read

Let's Look at Outer Space by Tim Furniss (Wayland, 1989)

Jump! Space Books (series) by Ian Graham (Two-Can, 1991)

On the Moon by Angela Grunsell (Franklin Watts, 1983)

How High Is The Sky? by Meredith Hooper (Simon and Schuster, 1990)

Stars and Planets (Usborne Young Scientist, 1991)

Spaceflight (Usborne Young Scientist, 1991)

Index

Notes for Adults

The content of this book is linked to areas of study outlined in Key Stages One and Two of the National Curriculum for Science.

There are many ways in which children can experience the wonder of space, the best starting point being actual 'star gazing'. Wrap up warmly and try to find an area away from house or street lights. A child's first view of the Moon through a pair of binoculars is a memorable one. Encourage children to talk about what they see and to hypothesise. You can help them in this by asking open-ended questions. Point out to them the phases of the Moon. Similarly, get them to notice where the sun rises and sets, and how its height in the sky changes between winter and summer.

Model-making is always an effective way to involve children. Designing and making their own model moon buggy with moving wheels would be a fun DT task. A simple demonstration involving spherical objects and a torch is also probably the best way to explain the movement of the planets, the phases of the Moon, the seasonal changes caused by the Earth's movement around the Sun etc.

In addition, there are many excellent science museums and planetariums which put on exhibitions and demonstrations especially for young children.

Places to Visit

Liverpool Planetarium
Liverpool Museum
William Brown Street
Liverpool
Tel. 051 207 0001

Science Museum
Exhibition Road
London SW7
Tel. 071 589 3456

London Planetarium
Marylebone Road
London NW1 5LR
Tel. 071 486 1121

Armagh Planetarium
of Northern Ireland
College Hill
Armagh
BT61 9DD
Tel. 0861 524725

Mills Observatory
and Planetarium
Balgay Park
Dundee
DD2 2UB
Tel. 0382 67138